The Kids' Career L

A Day in the Life of a
Firefighter

Mary Bowman-Kruhm
and Claudine G. Wirths

The Rosen Publishing Group's
PowerKids Press™
New York

Thanks to Jennifer Morimoto, David Swanson, and the Montgomery County, Maryland, Fire Department.

Published in 1997 by The Rosen Publishing Group, Inc.
29 East 21st Street, New York, NY 10010

First Edition

Book Design: Erin McKenna

Photo Illustrations: All photo illustrations by Kelly Hahn.

Bowman-Kruhm, Mary.
    A day in the life of a firefighter / by Mary Bowman-Kruhm and Claudine G. Wirths.
        p.   cm. — (The kids' career library)
    Includes index.
    Summary: Describes the daily responsibilities, tasks, and life of a firefighter.
    ISBN 0-8239-5094-8
    1. Fire extinction—Juvenile literature. [1. Fire extinction. 2. Firefighters. 3. Occupations.
    ] I. Wirths, Claudine G. II. Title. III. Series.
TH9148.B68 1997
628.9'25—dc21
                                                                    96-53129
                                                                        CIP
                                                                        AC

Manufactured in the United States of America

Contents

A Firefighter's Day Starts

Firefighter Jennifer Morimoto arrives at the station by 7:00 every morning. She works on the day shift at the fire station. She takes off her coat. Then she carefully puts her firefighter **gear** (GEER) near the fire truck. She is ready if the station gets a fire **alarm** (uh-LARM) call.

Some firefighters are already on duty when Firefighter Morimoto gets to the station. The night **crew** (CROO) leaves when the day crew arrives. The station always has some firefighters ready in case there is a fire.

◄ Firefighter Morimoto starts her day by collecting her firefighter gear.

Ready to Go

Next, Firefighter Morimoto checks that the fire trucks are filled with **fuel** (FEWL). She wants to be sure they are ready to go when the fire alarm rings.

Then Firefighter Morimoto exercises. She has to be strong enough to help all kinds of people out of a fire. She stretches, runs, and lifts **weights** (WAYTS) to stay in shape. When she finishes her exercises, Firefighter Morimoto puts on her blue **uniform** (YOO-nih-form). She will wear this underneath her firefighter gear.

Exercise helps Firefighter Morimoto ▶
to stay strong for her job.

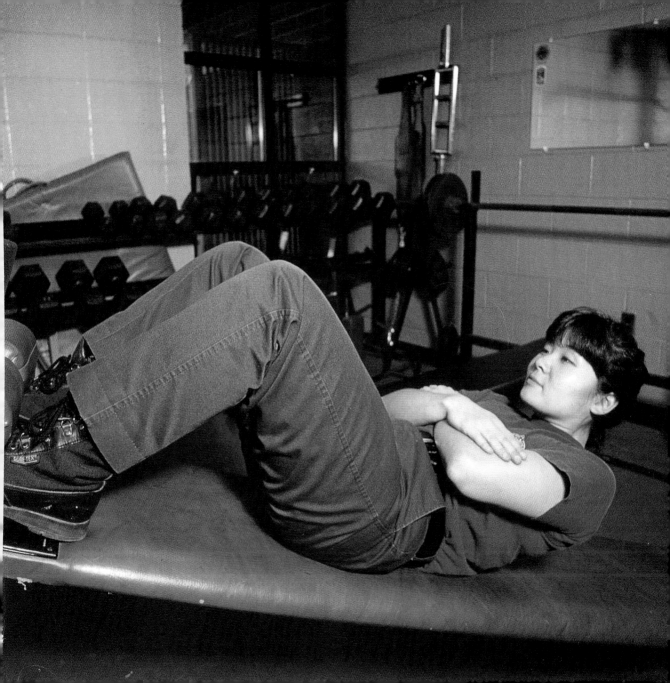

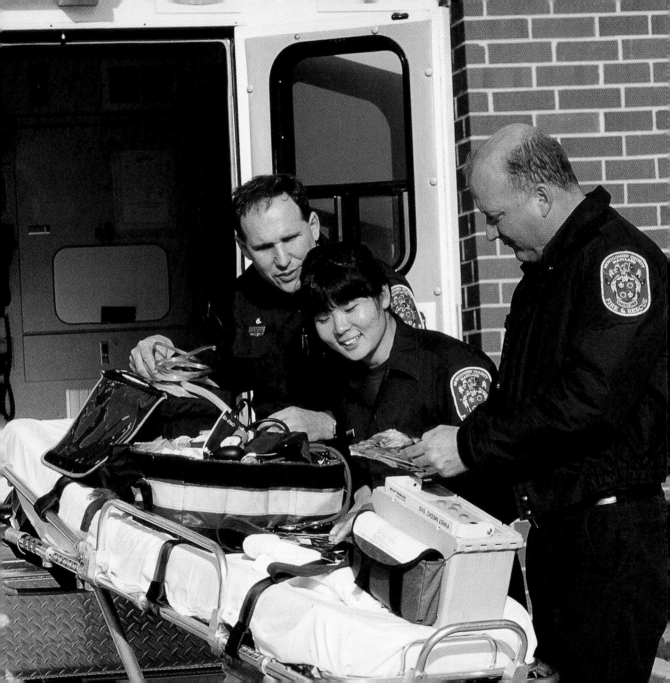

Training

Every morning, if there is no alarm, Firefighter Morimoto takes part in **training** (TRAY-ning). Training helps her do her job better. This morning she and the rest of the firefighters learn about a new **ambulance** (AM-byoo-lents) at their station. They are firefighters first, but they also have to know what to do if someone is hurt or sick. Firefighter Morimoto climbs inside the ambulance to learn how the controls work and to check the **supplies** (suh-PLYZ).

◄ Training is an important part of being a good firefighter.

Help!

Suddenly the fire alarm rings in the station. A person has called. There is a house on Lark Street that is on fire. The firefighters must get there fast!

Firefighter Morimoto runs to get her boots, pants, jacket, and **helmet** (HEL-met). All of her gear is in a pile so she can put it on quickly. She grabs a gas mask, too. If she has to go into heavy smoke, the mask will help her breathe.

Firefighter Morimoto is always ready to go if there is a fire. ▶

Lights and Sirens

Firefighter Morimoto turns on the engine of the fire truck. She turns on the lights and the sirens. Two more firefighters jump on the truck.

AH-WHOO! AH-WHOO! The siren tells people that the fire truck is coming. The truck leaves the station and heads for the fire. As it roars down the street, cars move to let it pass. Many boys and girls walking along the street stop to watch. They wish they could ride on the fire truck.

◀ The fire truck has bright lights and loud sirens that tell people to move out of the way.

At the Fire

When the firefighters get to the house, there is no fire. It was a **false** (FAWLS) alarm.

"Why do people make false alarms?" a firefighter asks.

"It's sad that some people think a false alarm is funny," Firefighter Morimoto says. "If there had been a real fire somewhere else, the people in that fire might have been hurt because we had gone to the false alarm."

Firefighters must answer all fire alarm calls, even if one turns out to be a false alarm. ▶

Writing a Report

Back at the station, Firefighter Morimoto writes a report on where the truck went and what the firefighters saw and did. She writes a report every time the truck goes out. Her report helps the firefighters learn more about fighting fires. This time it may also help the police catch the person who made the false alarm.

When she finishes, she helps the other firefighters wash the truck. They are proud of their trucks and keep them shiny and clean.

◀ Firefighter Morimoto keeps track of the alarm calls by writing reports.

Schoolchildren Visit

That afternoon, a class visits from a nearby school. Firefighter Morimoto tells the children what to do if their clothes catch on fire. She shows them a special fire safety rule called STOP, DROP, and ROLL. "Remember—

- STOP where you are,
- DROP to the ground, and
- ROLL around to put the fire out.

This rule could one day save your life," Firefighter Morimoto tells them.

STOP, DROP, and ROLL is an important fire safety rule that everyone should know. ▶

More Training

Later, Firefighter Morimoto and another firefighter work with a new hose. Pulling a heavy hose is hard so they help each other. They will help each other during a fire, too.

Water from hoses puts out a fire, but hoses help firefighters in another way. A house on fire is hot. And even on a sunny day, a house can be dark as night inside because of smoke. The firefighters cannot always see through the smoke so they follow the hose to help them find their way out of a dark, burning house.

◀ More than one person is needed to carry a heavy fire hose.

Home at Last?

At 5:00, Firefighter Morimoto's workday is over. She looks for her coat.

Suddenly the alarm bell in the station rings again. Another fire call! Once again she runs to the truck. The night crew has not yet come to work, so the day crew must do the job. But Firefighter Morimoto does not complain. "We are firefighters. We go to the fire no matter what time it is," she says proudly.

AH-WHOO! Away she goes again!

Glossary

alarm (uh-LARM) The loud bell that tells firefighters they are needed at a fire.

ambulance (AM-byoo-lents) A special truck for carrying people who are sick or hurt to the hospital.

crew (CROO) A team of people who work together to do a job.

false (FAWLS) Not real or not true.

fuel (FEWL) A liquid that makes an engine or motor run.

gear (GEER) Clothes or tools used to do a certain thing.

helmet (HEL-met) A covering that protects the head.

supplies (suh-PLYZ) Things a person needs to do a job.

training (TRAY-ning) Learning how to do a job well.

uniform (YOO-nih-form) Special clothes worn for a job.

weights (WAYTS) Heavy things that, when lifted regularly, make a person's body stronger.

Index